PROSPECTING FOR
LODE
GOLD

GREGORY V. STONE

Dorrance & Company • *Philadelphia*

To my wife and partner, Melody

CONTENTS

FIGURES

Introduction

I wrote this book because of utter frustration with the scarcity of information I could find in any one book covering the subject of gold prospecting. The current book market has produced nothing of true value to the amateur gold prospector looking for the basic, essential facts necessary for finding gold.

Five years ago when I first got interested in gold prospecting, I sent for numerous books and pamphlets that were advertised in various treasure magazines. Most of them dealt with gold panning and placer mining. None of these proved worthwhile. Because of this situation, I felt there was a need for a book such as this. This need has been magnified by the current prospecting boom brought on by the high price of gold. Also, I hate to see the big mining companies claiming all our state and federal lands.

All my information had to be obtained piecemeal through research of government mining publications, countless geology books, and by associating and talking with knowledgeable prospectors. From each source, I extracted only information pertaining to gold.

I am sure that anyone armed with a knowledge of the basic information necessary for gold prospecting, which I feel is set forth very adequately in this book, may very possibly find the "Mother Lode." This, however, will take time, many weekends and vacations, but it is worth the effort. Even if you do not find the "Mother Lode," I guarantee to you that prospecting

1

is good for both the spirit and body. In addition, it is good, healthful family recreation. Prospecting is somewhat like gambling; not everyone will win, but someone will, and it might be you.

Incidently, I have now prospected for five years and last summer was fortunate enough to stake my first lode claim, which I named after my partner and wife, Melody. You will find this book interesting and it will stir your blood. Good prospecting!

Chance of Finding Gold

Your chance of locating lode gold today is greater than what it was in the 1800s and early 1900s provided you go about it in a scientific way. The old-timers did not have four-wheel drive vehicles nor thousands of miles of roads which would take them to any mountain range they wanted to prospect. Do not let anyone tell you that all the gold has been found; it just is not the truth.

Let us face it. Most of the old-timers flocked to areas where gold was discovered. Countless areas have not been prospected adequately, and some areas were even overlooked. The early seekers mostly followed the streams. We do not have to. They did not have our superb government and state bureaus of mines to rely on, nor the countless geology and mineralogy books that, even though lacking in specific gold data, can be valuable sources of prospecting information.

The metal detector, which was nonexistent in bygone days, can be an extremely valuable tool provided you know its limitations. The old-timers got the greater share of the placer gold, river gold, but failed when it came to finding lode gold. Lode gold, the originating source of all gold deposits, has fed placers and alluvial deposits through decomposition and erosion, but much lode gold ore is still trapped underground, filling fissure veins in native rock. This is the gold that we are interested in. It has not eroded sufficiently enough to feed placers or alluvial deposits, but has only slightly outcropped, and must be searched out using scientific clues that are provided at or near the outcrop. This book will point out how to recognize these clues.

Researching for Potential Gold-Bearing Areas

Before we start on a prospecting outing, we must select an area that is likely to be mineralized. Regardless of the state in which you reside, it will have a Bureau of Mines that has published mining papers covering your state, which can be obtained from your city library. These publications can be checked for reference material. They will show maps of your respective state diagraming intrusive mountains along with symbol codes depicting what minerals and ores have been found in them. We are concerned primarily with intrusive, igneous formations because they contain lode gold. I will go into this later.

After locating the intrusive area that we wish to prospect, we must obtain a state map so that we may pinpoint the area selected. We will accomplish this by using a reference point, such as a river or town, so that we may plot a transportation route into the area selected. In some outings, we may have to leave the main highway and travel on dirt roads which pass through private lands in order to reach United States Forest Service land. In these cases, I suggest that you write the applicable forest service district office, requesting a forest service map of the area. These maps will show forest service access roads that pass through private land. Forest service access roads are usually the only way to reach inaccessible United States Forest Service lands. If you do not have a forest service map, you will think you are trespassing on private land because many farmers and ranchers place "No Trespassing" signs on fence posts in such a position so as to make you believe that you are trespassing and on the wrong road. However, you are not, and, as an American citizen, you have the right to use these access roads.

If you ever want to prospect on private land, be sure to obtain permission. Also, when passing through private land, be certain to close gates. If we respect the rights of the rancher and the farmer, he will not be so determined to keep us off county and forest service access roads.

Also, I might add, many historical books are available that describe old mining camps and towns and their locations. Let us face it. The early prospectors may have left some gold behind, but I believe in most cases they were pretty thorough. However, I must confess that the "Mother Lode" that supplies many rich placers was never found by the old-timers. Still, I would rather seek out an untapped area that has potential than search for overlooked treasure.

For our first trip, we will follow a river and its tributaries into the mountains.

CHAPTER THREE

Tools Required for Gold Prospecting

Now that we have selected our area, we will need the following prospector's tools. Since we will be prospecting up a mountain stream, a gold pan and a small collapsible shovel are absolute musts for taking gravel samples. I prefer using the twenty-inch, black plastic gold pan with built-in riffles and a small sturdy shovel that has convertible application and can substitute as a pick. Make sure the shovel is of high quality construction because there are many inferior ones on the market. Normally, these inferior shovels will break the first time they are used.

When we prospect away from the stream in search of the gold's source or "Mother Lode" we will need a miner's rock pick, a small five-pound sledge hammer, an 18-inch tempered steel punch that has been ground to a point, and a 3/8- to 1/2-inch star drill. Remember that when filing or grinding the end of your punch, do not let it get too hot because excessive heat will take the temper out of the steel. Select a good pair of heavy-duty leather work gloves, and use plastic goggles for protection from flying rock chips. One other item, if you are lucky enough to own one, is a metal detector, which can be a valuable tool for finding metallic ores or ore oxides that are near the surface. However, it cannot tell you how much or what you have found. Deep-seated deposits will not be detected unless you have a very expensive unit with superior penetration capabilities.

Now we need a pack frame and pack sack to carry our tools and to carry out the ore samples that we may find. Before departing on your trip, always tell a friend or relative where you will be prospecting and when you plan to return. This could save your life.

Gold Placer Deposits

Placer deposits are the results of decomposed and eroded mineral and metal-bearing rocks (lode gold) being carried by water and glaciers from higher hills and mountains, and the gold and materials they contain being deposited into basins, faults, creeks, and rivers. Placers are mostly found in river beds (Fig. 1). Therefore, as we proceed up the stream, we will take samples from the gravels at various spots within the stream bed and from gravels lying away from the water's edge.

Once you have loaded your pan with gravel, place it on the bottom of the stream, covering the pan with water. Then, with your hands, break up the dirt lumps, moving the pan vigorously back and forth, from side to side, allowing the flowing stream to carry away the muddy water. As the water in the pan clears, change from a shaking, back-and-forth motion to a gentle, swirling motion, rotating your pan counterclockwise to eliminate remaining muddy water. Take out the large rocks after making certain they are washed clean. Again, shake the pan back and forth and rotate remaining material, spilling out lighter materials and continuously looking for gold. Now, set the pan down and remove small rocks. Check for any pebbles or grains that are red, green, brown, blue, pink, transparent, or metallic in color. If you find these indicators, it shows that you are in an area of mineralization.

Now, place the pan with remaining concentrates back into the water. Tip it so that the front lip of the pan (farthest from you) is an inch or two lower than the rear lip (nearest to you).

FIGURE 1—STREAM OR
ALLUVIAL PLACER FORMATION

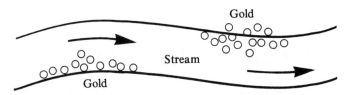

In a rapidly flowing, meandering stream, the fastest water is on
the outside curve of meanders and slow water on the inside curve.
The junction of fast and slow water, where gravel beds form, is the
area of gold deposition.

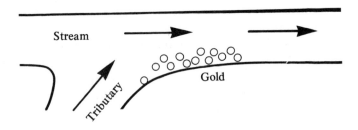

Where minerals and gravels are transported by a swift tributary
into a slower-flowing stream, the gold accumulates at the near
side.

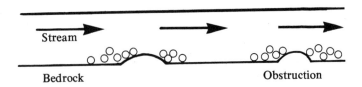

Where streams cross highly inclined rocks, the gold is caught, as in
the riffles in a sluice box.

Do not tilt the pan too much or you could lose the gold. Continue this routine until only the heavier material is left in your pan and again check for nuggets. If you have found only ordinary looking sand and gravel, you will assume you are just wasting your time in this area. However, if black sand, magnetite, an iron-bearing sand with a high specific gravity, is found lining your pan, it usually is a good sign, as it is usually associated with gold. Other good indicators of mineralization are garnets, olivine, zirconium, topaz, quartz crystals, and pyrites. Do not waste your time collecting flour gold. We are interested in nuggets and their source.

If we have found indicators showing that we are in a mineralized area, we can now progress upstream seeking bedrock. Bedrock is the solid rock underlying the overburden of a stream. In streams, gold will usually be found on the downstream side of obstructions, on the inside bend of curves, potholes or cracks in bedrock, and on the downstream side of the area where a tributary enters the stream. Coarse gold is found more often in deeper areas and fine gold in shallow areas. One thing that I should mention at this point is that angular-shaped gravel beds very rarely contain gold as this gravel has not traveled very far from its source. Gravels that are rounded, abundant with quartz pebbles, and, as I mentioned, showing magnetite (black sand), ilmenite (white sand), garnets, zircons, yellow sand, and monazite, should be gone over thoroughly. The most likely place to find gold is where bedrock is exposed on the bottom of the stream, or where bedrock can be reached easily by removing the overburden.

If we are lucky enough to spot cracks or a pothole on the bedrock's surface, we should clean out the cracks and holes and pan the collected material. If there is gold in the area, it will be deposited here. Gold usually lies on the bedrock or within a foot or two of it. If you cannot find bedrock or dig down to it easily, you will be wasting your time.

If rough gold flakes or nuggets are found in our pan, we can assume that we are close to the gold's source. The angularity of gold is inversely proportional to the distance traveled. If what you find looks like gold, take it from the pan and place it between your teeth. Gold will not crumble when you bite down, but is malleable and will show the indentation of your tooth. If it is fool's gold, pyrite, it will not indent when you bite it. Pyrite, iron sulphide, is usually cube-shaped, pale brassy yellow and has a metallic luster. Its physical appearance is nothing like gold. However, pyrites can contain gold between their planes of cleavage. Stream gold does not have cleavage; it takes on almost any conceivable shape. It looks like molten metal or a drop of solder.

Gold can range in color from silvery yellow to yellow, and silver white to orange red when impure. I have seen gold that is coated black. Gold does not rust, but will accept a rust or alloy coating of another mineral. This is rare in most areas, however. The black-coated gold that I saw came from a placer in Alaska. Some amateurs are also fooled by weathered mica, and pyrites of copper that look like flake or flour gold, but these, too, will crumble when struck with a miner's pick and, if thin enough, when squeezed between your fingers. Gold will flatten when hammered as it is extremely malleable. Also, I recommend the use of a little nitric acid for testing suspected gold nuggets, as both iron pyrites and copper pyrites (chalcopyrite) and mica are soluble in this, whereas gold is not affected.

Placer gold tends to occur in concentrated pay streaks that are narrow and relatively rich. If the gold that we found is coarse, we will proceed up the stream taking additional samples. If we find no more gold nuggets upstream, we will assume that we were in the area of the gold lode that fed the placer and will leave the stream bed to search for it.

Intrusive and Extrusive Rocks

Before we can discuss the various theories of ore deposits, we must understand the creation of an intrusive area, which I mentioned earlier, is the area in which you are most likely to find lode gold. This is because intrusive igneous bodies, formed by magma or igneous-rock-forming fluid, are usually the source of mineralizing solutions.

First, I will mention the batholith, which is the largest type of igneous, intrusive body. It is formed from the plutonic, or very deep rocks, that are believed to have formed near the boundary between the mantle rock and the earth's crust. Plutonic rocks are molten rocks of stupendous size. Large mountain chains, such as the Rockies, are but mere bulges of larger plutonic masses underneath. From them appear the batholith, a bulging, upward extension of the plutonic material underneath known as the magma, that has pushed its way to the surface by buckling and melting rocks above it.

The batholith has no known bottom and is up to 100 miles wide and several hundreds of miles long. Intrusive magmas usually push surrounding rock masses aside to such an extent that they are folded and altered considerably by the heat of the igneous mass. That is, the magma has literally pushed its way into the host or intruded rock by force and, in so doing, has induced the development of outer border features such as joints, fissures, dikes and veins. This is why most areas of igneous activity and its subsequent mineralization are associated with mountainous belts showing strong folding and faulting.

FIGURE 2—INTRUSIVE ROCKS

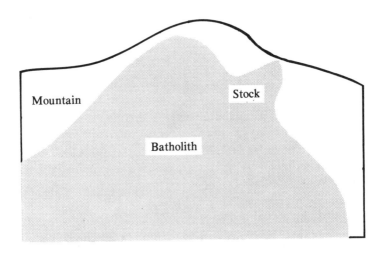

The cross section of batholith, the largest type of igneous intrusive rock, and its stock shows that the batholith forms the backbone of mountain chains and below it lies the magma or plutonic rocks. The stock is a large offshoot of a batholith. It is of smaller mass and has worked its way close to the surface before solidifying.

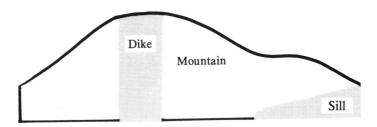

The cross section of a sill and dike shows that the sill, a part of the batholith, is a horizontal, or nearly horizontal, body of igneous rock lying between layers of intruded rock. The dike, another part of the batholith, is a nearly vertical, wall-like mass of igneous rock that cuts across intruded rock.

14

FIGURE 3—INTRUSIVE ROCKS

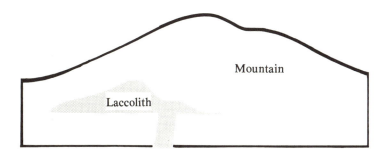

The cross section of laccolith shows that laccoliths are lenticular masses of igneous rocks lying between stratified rocks that wrap around them. They are part of the batholith.

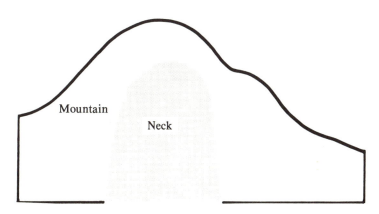

The cross section of volcanic neck shows that the neck is a column of magma that occupies a volcano conduit. It has solidified into igneous rock and is a part of the batholith.

Incidently, if you ever come upon limestones that have been intruded or lie near an intrusive area or are cut by fissures, it would be to your advantage to check out the area carefully for commercial ore deposits. Limestones are porous and can be good reservoirs for magmatic water solutions that may have borne minerals and ores.

Meanwhile, back to the subject of the batholith. Its crowns or tops are very irregular and project upward into the overlying rocks, creating gentle domes or elongated ceilings that are rounded. Once again, I will state that batholiths are igneous cores of magma that have been intruded into older mountain rock that already existed. Igneous bodies are formed directly from molten mineral matter, as were all rocks at one time or another. The batholiths and their dikes, sills, stocks, necks, and laccoliths (Figs. 2 and 3) are mineralized from the magmatic fluids that rise up from within the batholith and become concentrated in the crowns and ceilings of abovementioned rocks, or are intruded into the surrounding older formations that were already folded and uplifted. The mineral and ore-forming materials are carried in suspension or solution or as dissolved gasses of hot magmatic waters that flow upward into and throughout faults, fissures, porous rock and veins. We will go into the mineralization process in greater detail later on. As I mentioned before, from the batholiths are formed dikes, upright or steeply inclined or near vertical walls of igneous rock formed by magma that was forced into a break or crevice; sills, sheets of igneous rock that are forced nearly horizontally between two layers of strata bulging one layer of strata upward; stocks, large offshoots of a batholith being of smaller mass that have worked their way close to the surface before solidifying; necks, columns of hardened magma leading to volcanoes or to laccoliths. A laccolith is a dome-shaped mass of igneous rock that has been forced between older formations, most of which lie in layers or beds.

Before we go on, I will briefly go into extrusive rocks or lava rocks (Fig. 4) originating from volcanic activity. As I men-

FIGURE 4—EXTRUSIVE OR LAVA ROCKS

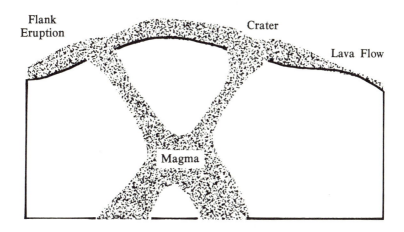

The cross section of extrusive or lava rocks shows that extrusive rocks are poured out upon the earth's surface and result when the magma cools quickly as its surface is exposed to the atmosphere.

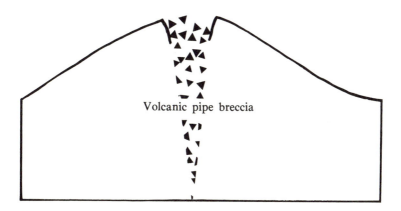

The cross section of volcanic pipe breccia shows the broken sharp-edged material confined in the solid walls left by volcanic action which drilled pipelike holes upward through the rock. These can provide ideal channelways for mineralizing solutions. Breccia deposits have yielded some gold.

17

tioned before, the intrusive rocks solidify within older rock masses, whereas extrusive rocks are poured out upon the earth's surface and their molten magma cools very quickly because its surfaces are exposed to the atmosphere which quickly dissipates the heat. The rapid cooling process results in rocks of very fine grain. The lava deposits normally had no chance to become mineralized or to carry rich ore deposits. This is because the violent eruptions offered less opportunity for a concentration of the metal-bearing emanations that normally flowed upward during a slow cooling process. Therefore, volcanic ore deposits are few in number and are of little economic importance.

However, in some instances, volcanic pipe breccia deposits have made ideal channel ways for mineralizing solutions and provided much open spaces for mineral deposits. Breccia is fragmented, sharp-edged materials confined within solid walls left by explosive volcanic activity which drilled pipelike holes upward through the rocks. Breccia remained, falling back into these pipes and has later been mineralized and possibly could yield gold and silver. The intrusive rocks remained beneath the surface and, therefore, did not cool as quickly as the extrusive rocks. Consequently, their grain size is generally larger than that of the extrusive rocks.

The intrusive rocks, stocks, and batholiths are only revealed at the earth's surface through erosion of the older cover rocks or through subsequent uplifts in the earth's surface. Mountains that are young in geological time have not weathered sufficiently to expose their mineral wealth through mineral-bearing outcrops. These mountains can usually only be prospected with sophisticated, metal-detecting devices. Big mining companies have metal-homing devices that can be used from aircraft flying over a suspected area of mineralization. If a metal reading is recorded, a team will go into the area and take core samples to ascertain what metals lie beneath the surface. However, the amateur prospector without financial backing must rely upon his ability to find an exposed lode outcrop, vein, gossan, or dike. This will be discussed in detail later on.

Mineralization Process

Now we will go into the mineralization process in greater detail. First, all minerals and commercial ores are formed through crystallization from magmas, chemical deposition from liquid solutions, deposition from gases and vapors, and metamorphism. In crystallization from magma, we have many elements in a dissociated state that begin to gradually group themselves into mineral molecules as the magma begins to cool and slowly crystallize to form the mineral particles of the resulting solid rock. However, gold lode deposits are formed mostly by hydrothermal solutions in intrusive rocks. Here we have chemical deposition of minerals and ores from hot solutions of magmatic origin. The hot waters, called hydrothermal solutions, are given off in the late stages of magmatic cooling, and they flow through openings such as faults, fissure veins, (Fig. 5) and porous rocks seeking areas of lower pressure. These hot waters carry vast quantities of mineral and ore matter in solution. As the temperature and pressure drop in the solutions in their ascent into cooler rock areas, the least soluble minerals and ores become precipitated or deposited first, and the most soluble ones later. These hydrothermal minerals and ores fill openings in rocks such as fissure veins, pore spaces, and breccia openings. But, the single most important method of deposition of minerals from hydrothermal solutions comes about from chemical reaction between wall rock and solutions. By this process, ore and gangue minerals take the exact place of surrounding rock minerals,

FIGURE 5—FAULTS

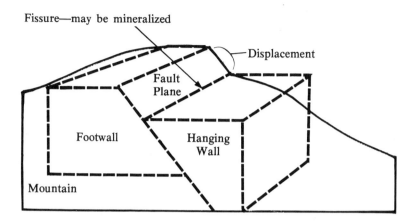

Fissure—may be mineralized

Displacement

Fault Plane

Footwall

Hanging Wall

Mountain

The cross section of a fault shows the displacement of rocks along a zone of fracture resulting in a fault. The fault surface may stand vertical, inclined or horizontal. Faults serve as channelways for mineralizing solutions.

the former being deposited from solution and the later being taken into solution with no change in volume. This is known as replacement, and this chemical transfer, molecule for molecule, gives rise to huge mineral and ore deposits. To most geologists, the hot water and the dissolved metals are simply igneous emanations from a cooling magma. Deposition from gasses and vapors is a less common mode of formation of crystallization. Metamorphism is the process whereby the constitution of any kind of rock that is subjected to enough pressure, heat, and chemical activity will change so that the original minerals will recrystallize and form new minerals. The most important changes during metamorphism take place as a result of hot vapors and hot fluids leaving the igneous masses and penetrating the surrounding rocks. Such vapors and fluids are often heavily charged with mineral

FIGURE 5— FISSURE VEINS

The cross section shows several fissure veins. Most fissure veins are narrow and range in length from a few hundred feet up to a few miles. Fissures are formed by stresses within the earth's crust and may or may not be accompanied by faulting. They may be mineralized by intrusion and its subsequent mineralizing solutions that flow through the veins.

matter. The large, hot igneous masses push themselves into older rock formations, making them intensely hot wherever the older rocks are in contact with the hot magma. The older rock's minerals are dissolved by heat and then replaced with new minerals entering via the igneous body. Needless to say, wherever you find an intrusive area, you will find metamorphic rocks. Metamorphism happens in sedimentary rocks (rocks formed by the accumulation of sediments) more frequently than in igneous rocks because the latter originally were formed from heat and pressure and thus are not likely to be changed much if subjected to heat and pressure again.

Metamorphic rocks are gneiss, slate, schist, serpentine, quartzite, and marble. The metamorphic rocks show distinct graininess, long thin streaks of minerals of contrasting color and foliation, or the formation of many thin sheets within the rock which can be easily split apart.

The majority of gneisses and schists are completely without interest to the prospector looking for commercial ore deposits. Gneisses rocks look like granite and their principal feature is parallel banding with streaked coloration, with or without fold or curves, due to buckling. Light-colored streaks or bands of felspar and quartz often alternate with dark streaks of mica or ferromagnesian minerals. Gneisses very rarely contain gold. Schists are fine grained, usually with very thin alternating layers rich and poor in mica. However, some visible gold has been observed in some crystalline schists. This is usually in small quantities. Quartzite is metamorphosed sandstone. Gold can be found in slates showing quartz veins (Fig. 6).

FIGURE 6—METAMORPHIC ROCKS

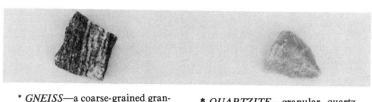

* GNEISS—a coarse-grained granular rock with irregular banding alternating with layers of darker minerals.

* QUARTZITE—granular quartz produced by metamorphism of pure sandstone.

* MARBLE—a metamorphosed limestone. When pure, marble is white.

* SERPENTINE—hydrous magnesium silicate.

SCHIST—may contain quartz, feldspar, mica, chlorite, tale, or hornblende. Some visible gold has been observed in crystalline schist.

SLATE—a fine-grained rock with perfect planar cleavage and enriched with mica. Slate sometimes has quartz veins that may contain gold.

* Relatively of little importance to the prospector.

23

Lode Gold Deposits

Ore deposits were settled at various temperatures and depths within the earth. Thus, we divide them into Epithermal, Mesothermal, and Hypothermal deposits. Epithermal deposits were formed in shallow vein deposits that were made at depths of less than 4,000 feet. Most of them formed within 1,500 feet of the surface at temperatures ranging from 50 to 200 degrees centigrade and the pressure scarcely exceeding 100 atmospheres. In these shallow vein deposits, banding is common and comb structure (Fig. 7), is more plentiful, and is much better developed than it is in veins of the intermediate zone. Examples of this group are found in gold and silver veins. Mesothermal deposits are found in intermediate depths of 4,000 to 12,000 feet. They formed at temperatures ranging from 175 to 300 degrees centigrade with a corresponding increase of pressure. The ore deposits themselves commonly show irregular banding with some crystals arranged like the teeth of a comb, pointing into the cavities. Their outcrops are exposed only by deep erosion, and they almost always appear in or close to intrusive bodies. An example of mesothermal deposits would be gold quartz veins. Hypothermal deposits are deep-seated veins which were formed near batholiths or other deep-seated intrusions by solutions whose temperatures ranged from 300 to 565 degrees centigrade, which is the crystallographic inversion point for quartz. The pressure was probably very high. Erosion and subsequent upheavals have brought

FIGURE 7—VEIN BANDING AND COMB STRUCTURE

Wall Rock

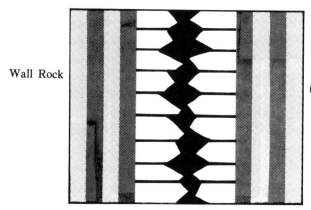

Wall Rock
(encasing vein)

☐ Quartz Crystals

▦ Clay Material (between minerals and ores)

▨ Blende (any of several minerals, such as sulfides)

■ Vug (small unfilled cavity in a lode vein)

many of these veins, which once lay 12,000 to 15,000 or more feet below the surface, to the exterior.

Hypothermal, gold quartz veins, together with pyrite and other sulphides and gold, were deposited from ascending mineral-bearing solutions. In the majority of hypothermal veins, the gold is so finely divided and uniformly distributed that its presence in the ore cannot be detected with the naked eye. This is why a prospector should always have his ore samples assayed.

To continue, most metalliferous, or metal-bearing deposits, are found in areas of igneous activity. Their habitat is predominately in regions where deep-seated igneous intrusives have cut through older-era rocks and their deposits are only exposed through deep erosion. These intrusive ore deposits are normally arranged zonally outward from a hot center or magmatic source. The higher-temperature or hypothermal ore minerals are formed and deposited nearest the igneous body's hot center, and the lower temperature or epithermal and mesothermal ore minerals are formed and deposited farther away and upward from the hot center. This process is called mineral zoning. The ore minerals will precipitate out of solution and go into crystallization or solid form when the temperature drops below their respective saturation points. Deposition of minerals is influenced by their specific gravity (weight of the mineral) and their solubility when in solution.

Gravity is an ever-present force tending to effect the required separation. A mineral may precipitate out at one level and sink downward to a lower level due to its higher specific gravity. This is why, in any one ore, the minerals have usually separated out in a certain well-defined succession, the later minerals in part replacing the earlier ones.

Veins are sheets of mineral ores that have been intruded into small faults, joints, and fissures. A fault is a fracture along which there has been relative displacement of the two walls. The fault may become a fissure vein by filling and replacement along its course. Jointing is a fracture in rock caused by the shrinking of a magma as it cools and crystallizes. A fissure is a continuous, tabular opening in rocks, generally of considerable length and depth. It is formed by compressive, tensile, or torsional forces operating on rocks and may or may not be accompanied by faulting. Fault surfaces may stand vertical, inclined, or horizontal although most are inclined.

In our search for gold, we will be mainly concerned with fissure veins. Normally, they are only a few feet in width, although some are considerably wider. In length, they usually range from a few hundred to a few thousand feet, but some have been known to extend a few miles. Their depth is measured in hundreds or thousands of feet. Few veins are vertical; most of them are highly inclined. The inclination is referred to as dip, which is measured as the degree of angle between the vein and the horizontal line of the rock. The horizontal course of a vein is referred to as the strike. Most veins curve gently along both the strike and dip, showing irregularity in width with pinches and swells. The outcrop of the lode is the intersection of the vein with the surface. Outcrop may be near the top, middle, or lower part of the vein. These veins are sometimes simple bodies, consisting essentially of one mineral, but normally are a composite made up of several layers. Sometimes these are all of the same mineral, though it may differ in color or texture. However, more commonly, composite veins consist of successive layers of different minerals and ores. The minerals most frequently encountered in veins are quartz and carbonates such as calcite, fluorite, dolomite, and limestone (Fig. 8).

Fissure veins constitute the source of a large part of the world's production of gold and silver. Gold is most frequently found in quartz veins chiefly as stringers (wire gold), thin leaves of plates and in scattered masses. Milky quartz, the predominant gangue mineral of gold, is colored from milk

27

FIGURE 8—MOST COMMON FISSURE VEIN MINERALS THAT MAY CONTAIN GOLD

CALCITE—Calcium carbonate. Crystals are hexagonal varying from tubular to needle-like. Worldwide occurrence.

FLUORITE—calcium fluoride, transparent to translucent mineral of different colors. Usually cubic crystals, also massive and fine-grained. Vitreous lustre. Found in sedimentary rocks, ore veins and pegmatites.

QUARTZ—Milky quartz shown. Silicon dioxide, hexagonal crystals and also massive. Milky quartz produces greatest share of gold.

DOLOMITE—formed from limestones, coral or marble by action of solutions containing magnesium. Worldwide occurrence.

LIMESTONE—sedimentary rock containing high percentage of calcium carbonate. Worldwide occurrence.

FIGURE 9—LESS COMMON FISSURE VEIN MINERALS THAT MAY CONTAIN GOLD

BARITE—colorless, white and light shades of blue, yellow, and red. Transparent to translucent. It occurs as a gangue mineral in metallic veins.

PYROXENE—Jadeite is shown. Other members of the pyroxene group include diopside, augite, aegirite and spodumene.

HORNBLENDE—Black or greenis black. Vitreous lustre. Worldwide occurrence.

FELDSPARS—Feldspars are potassium, sodium, calcium, or barium aluminous silicates. Worldwide as dominant or component of most igneous rocks.

28

white to snow white and is usually found in massive form and not in crystal form. Gold, however, is found in other gangue minerals and in large ore deposits.

Perhaps here I should explain the meaning of gangue and ore. An ore is the mixture of ore minerals and gangue, from which metals may be extracted at a profit. Gangue minerals are associated undesirable minerals of an ore deposit. Veins far away from the intrusion are mainly filled by barren gangue without ore minerals. Quartz is the most plentiful gangue mineral, although calcite, barite, and fluorite are common. Pyroxene, hornblende, feldspar and a variety of other minerals are also found as gangue (Fig. 9).

Metallogenic elements or "ore elements" which include those of economic importance are: tellurides, sulphides, selenides, and arsenides localized in mineral deposits. The gold telluride ores are calaverite, sylvanite, petzite and krennerite. Calaverite, gold telluride, is brass yellow to silver white, opaque, brittle, and has a metallic luster. Its prismatic crystals are long, slender, and striated, distinguishing them from pyrite. The lack of cleavage distinguishes it from sylvanite. Sylvanite, a telluride of gold and silver, is silver white, opaque, brittle, and has a metallic luster. Its good cleavage distinguishes it from calaverite. It is a rare mineral associated with calaverite and is usually found in veins formed at low temperature. However, it may be in higher-temperature veins. It outcrops in skeleton forms deposited on rock surfaces and resembles writing in appearance. Petzite, a silver gold telluride, cubic, normally in granular masses, steelgrey to iron black, has metallic luster and good cleavage in one direction. Krennerite is rare.

A considerable amount of gold is found in and recovered from quartz, a silicon dioxide; calcite, a calcium carbonate; siderite (synonym of hornblende) ferrous iron carbonate; native gold; anriferous pyrite, iron sulphide; anriferous chalcopyrite, copper iron sulphide, and smaller quantities from arsenopyrite, iron arsenide sulphide; pyrrhotite, iron sulphide and galena, lead sulphide (Fig. 10). Quartz is usually colorless or white, but frequently colored by various impurities. It is found in rocks associated chiefly with feldspar and muscovite and in veins with practically the entire range of vein minerals. Quartz is an important ore of gold. Calcite is one of the most common vein minerals, occurring as a gangue material with all sorts of metallic ores; siderite is found in sedimentary formations, ore veins, and petmatites where it is often associated with calcite, barite, and the sulphides. Pyrite sometimes contain enough gold to make it a useful ore of gold. Also, it is a widespread accessory, being especially predominant near mineralized zones or veins, although it is not restricted to their vicinity. Chalcopyrite, copper pyrites, arsenopyrites and pyrrhotite are often closely associated with gold and may carry gold and silver. Galena, which is a compound of lead and sulphur, sometimes contains both gold and silver. Also, I should mention electrum, which is the natural alloy of gold and silver. It is cubic, massive and pale yellow to yellowish white. Its gold contains unusually high percentages of silver, 20 to 40 percent.

The most important members of the igneous rock family are granite, syenite, diorite, gabbro, felsite, andesite, basalt, obsidian, scoria, and tuff. Of these rocks the syenites, diorites, andesites and basalts are sometimes the host rocks for gold deposits (Fig. 11). Syenite is a crystalline rock that resembles a light-colored granite in appearance with little or no quartz. Its dominant mineral is orthoclase, and it contains lesser amounts of plagiclase and some hornblende, pyroxene and biotite. Quartz veins intruded into syenite have produced gold. The syenite porphyries or porphyritic rocks are also the host rock for gold ores.

Porphyries (Fig. 12) are nothing more than large crystals within a matrix of very fine crystals. The larger crystals are called phenocrysts. These larger crystals, or phenocrysts, were formed as the magma was cooling at a slow rate. When the magma moved upwards toward the surface, it cooled rapidly

FIGURE 10—SULFIDE MINERALS
THAT MAY CONTAIN GOLD

PYRITE—iron sulphide, cubic crystals, also massive. Pale brassy—yellow, metallic lustre. Worldwide occurrence.

CHALCOPYRITE—copper iron sulphide. Brass-yellow or golden-yellow, frequently with iridescent tarnish. Metallic lustre. Worldwide occurrence.

GALENA—lead sulphide. Cubic, tabular crystals or massive. Cleavable, coarse to fine-granular. Lead-grey metallic lustre. Worldwide occurrence.

ARSENOPYRITE—Iron arsenide sulphide prismatic or short prismatic columnar crystals. Can be granular or compact. Lustre. Worldwide occurrence.

PYRRHOTITE—Iron sulphide hexagonal. Tabular or platy crystals or granular masses. Bronze-yellow to brown. Metallic lustre.

FIGURE 11—IGNEOUS ROCKS

GRANITE—a coarse-grained granular plutonic rock. Worldwide occurrence.

GABBRO—Igneous rock containing essentially plogioclase. Worldwide occurrence.

FELSITE—includes the dense, fine-grained rocks of all colors except dark grey, dark green or black. Worldwide occurrence.

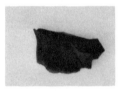

OBSIDIAN—Black natural glass.

SCORIA—Lava containing rough vesicles.

TUFF—rock composed of finer particles of volcanic ash and volcanic dust.

SYENITES—a granular rock of light color and even texture. It resembles granite in appearance. Sometimes the host rock for gold deposits.

DIORITES—a granular rock characterized by plagioclase feldspar. Sometimes the host rock for gold deposits.

ANDESITES—the volcanic equivalent of diorite and thus is composed chiefly of oligoclase or andesine feldspar. Sometimes the host rock for gold deposits.

BASALTS—a dark colored, fine grained volcanic equivalent of gabbro. Sometimes the host rock for gold deposits.

* Relatively of little importance to prospector.

FIGURE 12—PORPHYRY ROCKS

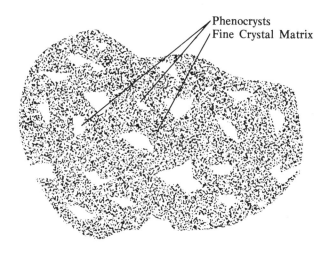

Phenocrysts
Fine Crystal Matrix

This cross section of porphyry rock shows light-colored phenocrysts in the darker groundmass.

as it came into contact with the cooler surface rocks. This
resulted in fine crystal matrix between the larger crystals.
Gold ore has also been found in limestones intruded by syenite
porphyry. Diorite hosts gold in many areas. It is composed
essentially of oligoclase or andestine and biotite, hornblend,
and, or pyroxene and is generally much darker in color than
granite but has the granitic look and texture. Andesites are
lavas that are mainly porphyritic, though some are compact in
structure. Thick flows are common in many places in the
western United States and some are noted for commercial
deposits of gold and silver. It is a hardened lava and may be
of pale gray or a red color. Usually the ores are in the veins of
quartz deposited in fissures. These veins are of a later origin
than the andesites, which have been altered by the ascending

hot solutions bearing the gold. The basalts are a fine-grained volcanic rock consisting predominantly of plagioclase and pyroxene, whose colors range from dark gray to green, purple and black. Most basalts are black, heavy rocks found in lava flows and sometimes on the margins of minor intrusions. Some basalts have been the host rock for gold deposits. Granites and granodorites, found in mineralized contact-metamorphic deposits of carbonate or dolomite rocks, are the source of many types of ore deposits, but rarely do they yield gold. Of the metamorphic rocks, slates and schists sometimes are host rock for gold deposits. Schists are fine-grained sediments with very thin alternating layers rich and poor in mica. Visible gold has been found in some crystalline schists. Slate is shale that is turned into slate through metamorphism. Through metamorphism, the clays of the shale deposits are changed to mica. Small mica flakes have grown along new cleavage surfaces to give the slate a luster. Gold has been found in quartz veins cutting between layers of slate with formation during the intrusive period, the quartz and gold being intruded into the older rocks of shale turning it into slate through metamorphism.

CHAPTER EIGHT

Prospector's Clues to Gold

Regardless of whether you are prospecting up a stream or in an area void of water, your methods of locating the lode will be essentially the same. Before we describe these methods, I would like to mention ancient river channels since we could very possibly come across one while prospecting.

At times, a modern river or stream will cross an ancient river channel that flowed in a different direction, and gold may have been redeposited from the gravels of the ancient river. Therefore, the ancient channel should be checked in both directions for a dry placer and then followed to the lode. Ancient rivers that have changed their course while cutting through deep river canyons have left gravel ledges high up on the sides of the canyon with possible ancient gold deposits. Also, placers have been found on hills where an ancient river bed was raised due to faulting. Never overlook a suspected ancient placer. If you live near a desert or rainless area, always be on the lookout for an Eolian Placer. Here, gold is deposited in a placer by wind, which has been the weathering agent instead of water.

Once away from the stream, we search for the "Mother Lode," looking for the telltale signs of mineralization. As I mentioned, any area of mineralization, even those void of water, will yield the same clues. Our clues will be Eluvial Placers, dikes, and gossans (Fig. 13), and veins. As we prospect, we must constantly search the terrain for anything out of the ordinary such as rock features showing dikes,

FIGURE 13—ELUVIAL PLACER

Erosion

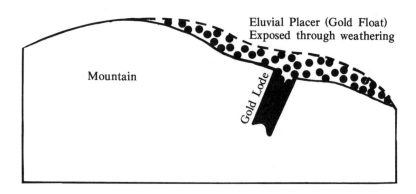

The cross section of an eluvial placer shows that the placer is formed without stream action; the materials and gold are released from weathered lodes that outcrop.

GOSSANS

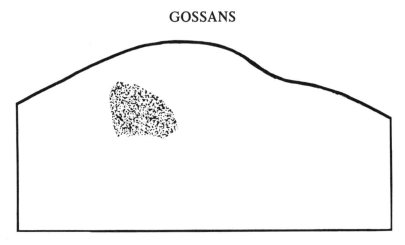

A gossan is an outcrop of a sulphide vein observed by noting iron oxides that stain the side of the mountain in colors ranging from orange to seal brown and maroon.

fissures, joints, and different coloration features within the same mass of rock. Also, by observing the rock fragments that we find on the ground, we should look for gangue minerals that are associated with gold.

If we find quartz fragments scattered on the ground, we would look for the source of the quartz. Who knows? We may have found an eluvial placer, which is the embryonic stage in the formation of a stream placer. Eluvials are formed without stream action; the materials are released from weathered lodes that outcrop above them. And, if you are lucky, the quartz outcrop may contain gold. Another clue to mineralization is pegmatites. A pegmatite is a variety of granite of coarse texture that occurs only in dikes, veins, and irregular pockets, usually deposited in schists surrounding granite intrusions, but it may be in the granites themselves. Pegmatites or pegmatite dikes normally are void of gold, but for some unexplainable reason, there are usually quartz veins in the immediate area that may host commercial gold deposits.

Granites that have weathered and fallen apart quickly indicate an area of strong mineralization and should reveal pegmatite bodies, and, as I mentioned, quartz veins. The granites in a mineralized area are often very coarsely grained and will crumble easily, whereas more durable granites usually indicate much less mineralization potential. The rapid erosive process of highly mineralized rock will result in rounding the granite and development of much surface debris. Once you know what to look for, the hills and mountains most likely to be mineralized will stand out against rugged cliffs of barren rocks.

Pegmatites also contain orthoclase, quartz, feldspar, microcline, microperthite and many possible ferromagnesian constituents in addition to the above-mentioned gems, and are formed by emanations from grantic magmas that have solidified in dike-like bodies in the enclosing rocks. A few pegmatites are found in limestone and others in sedimentary rocks. When searching for pegmatites, observe granite, schists, and gneisses rocks for marked changes in rock textures and coloration differences which may be due to pegmatites. Some large pegmatites can be observed as white streaks contrasting to the darker color of the enclosing gneisses and schists. Also check crevices and openings which appear to be different from ordinary joints. Look for trees and shrubs apparently growing out of solid rock as they require crevices for roots. Pegmatites that are completely or partially buried by over-burden can be located by finding float material such as quartz, large flakes of mica, feldspar crystals, and tourmaline crystals for they lie over or indicate closeness to a pegmatite body.

In level terrain, float fragments will be directly over the pegmatite from which they eroded, and, if you happen to be on a hill, be certain to make some allowance for downward movement of the float material. If one pegmatite is found, it is almost certain other pegmatites are in the immediate area. In addition, quartz veins are often found near pegmatites, and we will be most interested in them. Many pegmatite dikes turned into quartz veins which represent the last deposits of the solutions after they left the magma. Quartz, as we mentioned before, often carries gold and becomes an important ore of that metal. The largest pegmatites are found in schists and gneisses which fringe batholiths. Pegmatites range in size from twelve inches in thickness and tens of feet in length to perhaps twenty feet in thickness to several hundred feet in length. Pegmatites containing metals form in the crown and roof of the batholith. Sulphide minerals are rare in pegmatites.

In our search for gold, we will also be on the lookout for gossans. A gossan is an outcrop of a sulphide vein observable by noting iron oxides such as limonite and gangue materials. The limonite was formed during the oxidation of iron-bearing minerals, generally sulphides, which may yield valuable minerals. Colors of limonite are due to the mineral composition and are generally seal brown and maroon. Orange colors in the gossan capping or outcrop would indicate copper, and yellow and red colors would indicate pyrites, which may contain gold.

Another example would be a massive milky quartz vein that shows a rusty outcrop coating, the rust coming from weathered pyrite. It should definitely be examined for gold. Gold is commonly contained in base metal sulphides such as pyrites, galena, chalcopyrite, and related gangue minerals. However, gossans by themselves prove only one fact: that considerable quantities of iron and sulphides were decomposed to form the capping. Whether gold or silver minerals are present in depth remains to be proven by sinking a shaft and obtaining rock samples for assay.

The old-timers searched for rusty stains to locate sulphide veins, particularly quartz veins in which sulphides are commonly found with gold. Cavities within liminote are the places where gold is apt to be present in visible form. Fresh quartz should be removed from the vein with a miner's pick and observed for native gold or the telltale dark gray sulphides which often fill fissure quartz.

As we prospect, we must also be constantly on the watch for possible gold-bearing veins that outcrop on the surface of rocks or those veins protruding through the ground. Also, watch for rocks that normally host gold deposits such as syenite porphyry, diorite, andesites, basalts, some of the crystalline schists and slates showing quartz and intruded limestone rocks. Remember when prospecting, binoculars can be your best friend for scrutinizing the terrain.

The most common gangue minerals in vein outcrops are quartz, calcite, and fluorite. Quartz is usually the predominant vein filling in gold regions. Calcite is a common gangue mineral in many kinds of hydrothermal veins but is unlikely to be found in granitic-type rocks and pegmatites. Calcite gangue sometimes will be replaced by a later quartz gangue containing native gold. Fluorite is a primary gangue mineral of ore deposits and has been associated with high-grade gold ores. Fluorite is the synonym of fluorspar.

Other minerals commonly associated with gold are barite, quartz, galena, and sphalerite. Just because we locate a vein having one of the preceding gangue minerals, however, does not necessarily mean we will find gold or any other metal ore for that matter. Why some mineralizing solutions contain gold and others practically no metals is an unsolved mystery.

Assaying Your Ore

Once we have discovered a pegmatite, gossan, fissure, vein or likely host rock for gold, the only way we will know if we have found gold or any other metal is by taking mineral or rock samples and having them assayed for metal content. For example, the first samples taken from a vein should be selected from a depth of two to four feet. If you happen to find a rich outcrop, you may be able to spot native gold in your samples. However, most gold deposits in ores are microscopic and cannot be seen by the naked eye. An early assay will save you much exploratory work by letting you know if you should continue in your mining operation. For instance, if your assay came back negative, you might be wasting your time. If erosion in the area is slight, you may not have gone deep enough to expose the ore. However, if the erosion is great, you probably are wasting your time and should move on for prospecting elsewhere. If the assay came back showing a trace of gold and a small percentage of silver, you would want to dig deeper into the vein to see if the trace would become a percenage figure, and if the percentage of silver would increase with the next assay.

Gold and silver assays are quoted in ounces per ton, and an assay costs around $5.50. For your protection, the assay office will keep results and reports of your assay in strict confidence. Your State Bureau of Mines can tell you where the assay offices in your state are located. Also, your state school of mines may do your assay work at no charge or for a nominal fee. The average tenor of gold ore is .2 to .3 ounces per ton. Tenor is the metal content of an ore, which is expressed in ounces per ton for precious metals. The tenor varies with the price of the metal; the higher the price of a metal, the lower the metal content necessary to make it profitable. With the high price of gold now, many mines that were unprofitable will once again be profitable.

The value of the recoverable gold available in the ore must be greater than the cost of its extraction. Profit depends upon the amount of ore and the price of the metal and upon the cost of mining, treating, transporting, and marketing the metal. I believe access to the ore body is the most important single factor. Can trucks be moved in and out of the area to be mined without costly construction? Mining costs, such as having to build a road, can make a fairly rich ore deposit uneconomical.

CHAPTER TEN

Staking Your Claim

If we make a decision based upon our assay to proceed with the mining operation, our next step will be to stake, or file a mining claim. The first requirement by law is that a discovery must be made. No mining claim is valid until there has been a discovery of valuable minerals which would justify a person's expending time and money in the hope of developing a profitable mine. We have met this requirement by having a favorable assay report.

Our first step to filing a mining claim will be to determine exactly where our claim is located by using a suitable map and to tie at least one corner of the mine to a section corner, quarter corner, or a township corner of the public land survey providing that the area has been surveyed. This practice is not required by law, but current regulations do require that the claim or mine must be easily located from written directions (metes and bounds description), usually using a prominent land feature as a reference point. The land office can tell you if the area has been surveyed and can provide you with a record of the survey at a nominal fee. Also, the National Forest Service for the area in which you are mining can supply you with a forest service map showing sections and townships, if they are available. These survey maps will assist you to pinpoint your mine on the map, and enable you to tie it to a township corner. Referring your mine's location to a prominent landmark in the immediate area is an especially good idea if the area has not been surveyed. If you are having difficulties, the nearest Bureau of Land Management or National Forest Service can provide you with assistance. Also, there are several mining pamphlets available dealing with Federal Mining Laws and procedures for staking a claim.

The second step after locating the claim is to post a Notice of Location at the discovery pit or shaft. This will contain the names of the locators, date of location, and the approximate dimensions of the area claimed, plus its geographical location. Printed forms of location notices can be obtained from most print shops. I advise you to place the notice in a glass jar or metal can and to affix it to some permanent object at or very near the discovery shaft or pit. It is a good idea to have some disinterested party witness the posting and the signing of the notice. Within thirty days after posting the notice at the claim, you must distinctly mark your mine's boundaries on the ground so that it can be readily traced. You will need a compass when tracing and recording the direction of the boundaries.

To set up your boundaries, I suggest using four fence posts, four by four inches square by four feet six inches in length, set one foot in the ground at each corner of the claim. The corner posts must be marked with the name of the claim, and each post must have a number designation such as post, one, two, three, and four.

Your claim shall not exceed 1,500 feet in length along the vein and shall not extend more than 300 feet on each side of the middle of the vein. The maximum size as stated is 600 feet wide by 1,500 feet long. It may, however, be shorter or narrower, but who wants to take the chance on a claim that is smaller than allowed? Later, you may wish to stake additional claims. Before setting out the boundary posts, try to determine exactly how the vein runs. Trace it to the best of your ability. After posting the notice of location at the claim site, you have sixty days in which to do the discovery work to expose your ore or vein.

The third step is to record your location in the office of the County Clerk of the county in which your mining claim is situated. This may be accomplished after the boundaries are in place and recorded. It is necessary to have the Certificate of Location (note I did not say Notice of Location) verified before a notary prior to being filed. This form can also be obtained from most print shops.

From then on, you will be required to do annual assessment work. Not less than $100 worth of labor must be performed or improvements made annually. I suggest that you obtain a copy of your state's Mining Law Publication which is put out by the Bureau of Mines and Geology in your state for a small fee of about $2. It is a valuable tool and will include the required forms used when filing a mining claim.

If you are prospecting or mining on state-owned lands, you must apply to the State Department of Lands and Investments for a leasing permit. On private lands, you must obtain permission from the owner and you should have an agreement written on how profits will be split before any mining venture commences. Federal Mining Laws do not apply in these situations.

Dynamiting

If we are proceeding with our mining venture and plan to sink our exploratory shaft deeper into the vein, we will want to use dynamite. Without the aid of explosives, we would be wasting many valuable manhours by using a shovel, pick, and punch. Before you can use dynamite, however, you will be required to obtain a "Small Miner Exclusion Statement" from your Department of State Lands. Check with your particular state's requirement as they may vary from state to state. They will be sent at no charge. I strongly recommend obtaining a copy of a book entitled "Blaster's Handbook" prior to dynamiting. It is published by the Dupont Chemical Company and can be obtained in most public libraries. It is very well written and will answer most of your questions as to how to use explosives. Dynamiting is not difficult. Once you decide on what day you will dynamite, contact the head ranger of the district in which your mine is located for his approval. He will probably ask you to use electrical detonation caps because there is less chance of starting a forest fire with them. Also, notifying the ranger of your plans will enable him to alert his rangers. This will avoid their possibly reporting a forest fire when the blast occurs. It will send dirt and rocks into the air, which could be mistaken for smoke.

Other equipment required is plenty of electrical tape, ¼-inch wooden dowel whittled to a point for making a hole in the center of the dynamite stick to accommodate the electrical cap, one hard hat (or motorcycle helmet) for safety from falling rocks, one danger-warning sign mounted on a 5-foot post, one 250-to-500-feet roll of good electrical wire, two five-gallon cans of water if you plan on using mud pack, one roll of twine, one 10-foot 2-inch diameter pole for tamping, one 1½-inch star drill if you plan on hand boring dynamite holes, one large pail for mixing mud, one twelve-volt camping lantern battery, several sheets of blank cardboard paper, a few thumb tacks, and a marking pencil.

Always carry your dynamite separated from the electrical blasting caps. I package blasting caps individually in packing material and put them in a cardboard box well away from the dynamite. Check your state laws for transporting explosives. Dynamite cannot be stored in your home or within the city limits, so always plan on picking up the explosives from the magazine on the day that you will be using them. Never take more explosives than you can use in one day. After you have finished blasting, always make sure that all the sticks of dynamite and blasting caps have been used. Never leave dynamite or electrical caps behind. I find that I can work ten to fourteen dynamite sticks plus an equal number of caps a day. I suggest the first time you use dynamite you mud pack it. It is not as effective as boring, but it can be accomplished more quickly. Dynamite handled with caution is not nearly as dangerous as people would have you believe.

Once you get to the mine, carry the blasting caps and dynamite separately. Separate them on the ground. Since we are mud packing, I suggest starting off by firing one stick of dynamite at a time until you get used to handling it. Before starting, place your danger sign at the opposite end of the trail from where you will be detonating, or at a point where it would be possible for hikers or anyone else in the area to see. Write a message on the blank cardboard stating that you are blasting and warning the reader not to come forward until he hears the blast. Place the sign well away from the blasting area. This is a precaution that should always be taken. Before each blast, always check to see that no one is in the area.

47

Now we can lay out the roll of wire. I usually use 250 feet of wire providing there are trees or some sort of embankment that can afford protection from the blast. Always make certain the bare wires that will be touched to the battery are wound together and taped, making a closed circuit. After each blast, rewire them together and retape them.

Now, mix up clay-type dirt with a small amount of water in your mixing pail. Make a thick clay-type mud for best results. Take your stick of dynamite and push a hole into the center of the dynamite with the wooden dowel. Make the hole approximately one inch deeper than the length of the cap. Never use any metal device such as a knife, a pen, or a pencil for putting holes into the dynamite and do not cut it. Use electrical detonating caps with twelve feet of connecting wire. Take the cap and unwind the attached twelve-foot wires, laying them out straight on the ground. Next, place the cap into the dynamite stick. Once it is pushed into place, press the top of the dynamite stick together, sealing the hole. You may tape or tie the top of the stick if you wish, making certain the cap does not come out. Refer to "Blaster's Handbook" for methods of securing priming caps into the dynamite; half-hitching the leg wires around the primer cartridge is the easiest way.

If you already have a deep hole, you should lower the dynamite by tying twine around the center of the stick and placing tape around the twine so that it will not slip. Lower the dynamite into the pit and position it with your ten-foot pole. Gently place the mud over the dynamite, watching that you do not damage the wire coming out from the detonating cap. Once this is accomplished, tamp the mud pack solid with the ten-foot pole. Make sure the cap wires are not pinched or smashed. Now take the blasting cap wires and attach them to the wires of the 250 foot roll of wire. The blasting cap wires are enclosed in a casing for safety and should be attached one at a time, leaving one enclosed. Attach one blasting cap wire with one roll wire by winding them together, then wrap the bare wires with the electrical tape. Pull the other blasting cap wires out of its casing and attach it to the other end of the roll wire. You are now ready to blast. Go back to your safe area and separate the two wires at the end of the roll. Touch them to the battery posts. Off it goes! Always wear your helmet and look upward when you set off the blast. If rocks are sailing through the air, you want to see them so that you can get out of the way.

After the blast, again attach the two wires and retape them. Every effort should be taken to prevent misfires. In the event that the dynamite does not fire, do not return within thirty minutes of the time of applying the blasting current. Under most circumstances, the safest way to handle a misfire is to blast it right away. No attempt should be made to recover any part of a misfired charge.

It is usually unnecessary to remove the stemming or mud packing over the explosive as a new primer will ignite the original charge even though it is separated by considerable packing or steming. In other words, place a primer, either in a stick of dynamite or by itself, on top of your mud pack and fire it. It will ignite the unfired charge.

When you handle explosives, remember that you are dealing with potentially hazardous materials. Properly treated, dynamite is safe, but in inexperienced hands, it can be dangerous. Never, under any circumstances, should you ever do any blasting during an electrical storm. Suspend your blasting whenever a storm approaches. Get good counseling before you use explosives. Collect your best ore for shipment to the smelter.

Whether To Sell or Operate Your Claim

If you find that you have a good producing mine, you will have to decide whether to operate it on your own or to sell it to a big mining company. If you decide to do the latter, I suggest that you contact more than one mining company. Your local library should have a list of mining companies. In other words, do not take the first offer, but shop around. Who knows? You might receive a substantial amount of cash plus royalties, or a large block of the company's stock. If you decide to operate your mine on your own, I am certain a good assay report will find you an investor for your mining venture. Whichever way you decide, be sure you have a signed and acknowledged agreement between the company or other person and yourself.

GOOD LUCK